I0471926

Multi-Agency Ocean Rescue
Disaster Plan and Drill
Broward County, Florida

Investigated by: Reade Bush
J. Gordon Routley

This is Report 079 of the Major Fires Investigation Project conducted by Varley-Campbell and Associates, Inc./TriData Corporation under contract EMW-94-C-4423 to the United States Fire Administration, Federal Emergency Management Agency.

FEMA

Department of Homeland Security
United States Fire Administration
National Fire Data Center

U.S. Fire Administration Fire Investigations Program

The U.S. Fire Administration develops reports on selected major fires throughout the country. The fires usually involve multiple deaths or a large loss of property. But the primary criterion for deciding to do a report is whether it will result in significant "lessons learned." In some cases these lessons bring to light new knowledge about fire--the effect of building construction or contents, human behavior in fire, etc. In other cases, the lessons are not new but are serious enough to highlight once again, with yet another fire tragedy report. In some cases, special reports are developed to discuss events, drills, or new technologies which are of interest to the fire service.

The reports are sent to fire magazines and are distributed at National and Regional fire meetings. The International Association of Fire Chiefs assists the USFA in disseminating the findings throughout the fire service. On a continuing basis the reports are available on request from the USFA; announcements of their availability are published widely in fire journals and newsletters.

This body of work provides detailed information on the nature of the fire problem for policymakers who must decide on allocations of resources between fire and other pressing problems, and within the fire service to improve codes and code enforcement, training, public fire education, building technology, and other related areas.

The Fire Administration, which has no regulatory authority, sends an experienced fire investigator into a community after a major incident only after having conferred with the local fire authorities to insure that the assistance and presence of the USFA would be supportive and would in no way interfere with any review of the incident they are themselves conducting. The intent is not to arrive during the event or even immediately after, but rather after the dust settles, so that a complete and objective review of all the important aspects of the incident can be made. Local authorities review the USFA's report while it is in draft. The USFA investigator or team is available to local authorities should they wish to request technical assistance for their own investigation.

This report and its recommendations were developed by USFA staff and by TriData Corporation, Arlington, Virginia, its staff and consultants, who are under contract to assist the USFA in carrying out the Fire Reports Program.

The USFA greatly appreciates the cooperation from Deputy Chief Robert Bollia of the Broward County Fire and Rescue Division and Walter Houghton, Assistant to the Director of the Broward County Aviation Department in preparing this report.

For additional copies of this report write to the U.S. Fire Administration, 16825 South Seton Avenue, Emmitsburg, Maryland 21727. The report is available on the USFA Web site at http://www.usfa.dhs.gov/

U.S. Fire Administration

Mission Statement

As an entity of the Department of Homeland Security, the mission of the USFA is to reduce life and economic losses due to fire and related emergencies, through leadership, advocacy, coordination, and support. We serve the Nation independently, in coordination with other Federal agencies, and in partnership with fire protection and emergency service communities. With a commitment to excellence, we provide public education, training, technology, and data initiatives.

TABLE OF CONTENTS

Multi-Agency Ocean Rescue
Disaster Plan and Drill

Investigated by: Reade Bush

Local Contact: Deputy Chief Robert Bollia
                      Broward County Fire and Rescue
                      Division
                      2601 West Broward Boulevard
                      Broward, Florida 33312
                      (305) 831-8258

                      Walt Houghton
                      Assistant to the Director
                      Broward County Aviation
                      Department
                      1400 Lee Wagener Boulevard
                      Fort Lauderdale, FL 33315
                      (305) 359-6124

INTRODUCTION

On December 6, 1994, the Broward County, Florida, Fire and Rescue Division and a host of other local, state, and federal agencies participated in an aircraft disaster drill in the intracoastal waterway near the Fort Lauderdale/Hollywood International Airport. The purpose of the drill was to test the interagency response plan for a downed aircraft in the water. The plan was devised to integrate the response capabilities of multiple agencies under a unified command system.

Each of the agencies participating in the drill was charged with a specific emergency response role, and each brought different rescue capabilities which would be necessary to efficiently and expediently handle an off-shore crash with multiple victims.

The drill was designed to be as realistic as possible. It included hundreds of victims both on land and in the water, mutual aid fire and emergency medical agencies, the airport fire department, private ambulances, boats, helicopters, the media, and hospitals.

1

This report discusses the Broward County off-shore emergency response plan, the background considerations made that were used to set up the drill, the drill itself, and lessons learned from the drill. Many fire and rescue organizations can benefit from the concepts in this response plan, even though the plan was specifically designed for an off-shore response at a particular location. Its concept and design can be applied to almost any type of disaster that involves multiple agencies. The lessons learned are applicable to almost any type of multi-casualty incident or exercise.

BACKGROUND OF DRILL

There were several reasons for holding this drill. The Federal Aviation Administration requires large airports to conduct a comprehensive disaster drill at least once every three years. The drill must incorporate the potential agencies and jurisdictions which could be called to respond to a real emergency.

The Broward County Fire Department, the Coast Guard, and other agencies had never previously established a formal, unified command structure, incorporating all of the agencies into one plan. The airport's proximity to water bodies and the Atlantic Ocean creates a high probability that an aircraft could go down into the water. The purpose of the drill was to test the unified response plan and to familiarize rescuers from the various agencies with one another's capabilities.

THE PLAN

The plan, called the *Broward County Ocean Rescue Mass Casualty Plan* delineates the structure of operations at an off-shore disaster with multiple victims. It identifies the participating emergency agencies, responsibilities for each responding agency, jurisdiction and who is in charge of each function, communications, and the levels of response by agencies (see Appendix A for a copy of the plan). The mass casualty plan is designed for use on any ocean based disaster, including downed aircraft, ship fires, or shipwrecks. It employs the standard Incident Command System structure as a means of organization. This includes an Incident Commander who leads four major sections: Operations, Logistics, Planning, and Finance.

The purpose of the plan is "to provide an aircraft ocean rescue mass casualty incident command system at Port Everglades (a port area with easy access by emergency responders to the Atlantic Ocean and intracoastal waterway) near the airport, and to assign functional responsibilities to responding agencies[1]." The agencies function under the Florida statewide mutual aid plan (see Appendix B).

The plan is designed so that it could be implemented for either an incident in the intracoastal waterway or in the ocean. It recognizes that no one agency could handle a disaster the size of a commercial airline crash all by itself. For example, the fire department would need the resources of the Coast Guard to reach and retrieve victims. The Coast Guard would need the assistance of the fire department with treating and transporting patients to hospitals once they were brought to shore. The hospitals might need the assistance of the fire department's hazardous materials team to decontaminate victims. The plan attempts to coordinate the interaction that would occur as the agencies combined forces to handle a multiple victim plane crash.

Since the Broward County Fire and Rescue Division does not have boats to retrieve victims at sea, the Coast Guard would be called to perform this function. Additionally, helicopters from the Broward County Sheriff's Office, the Coast Guard, and the local Air Force base could assist with victim retrieval.

---

[1] Broward County Ocean Rescue Mass Casualty Plan, section 1.1.

Concept of the Plan--The plan's concept is to have the first arriving fire department units establish the command system on shore. The Fire Department would respond to one of two pre-designated shore locations, and a Battalion Chief would assume command. The Coast Guard and other assisting agencies would be notified of the incident by the Broward County Emergency Communications Center.

This design may seem unusual because the fire department establishes initial command of an incident that could be miles off-shore; however, by establishing command on shore, the fire department sets up a structure for receiving patients who will eventually be transported to shore.

The Coast Guard would dispatch boats and aircraft to the scene and its duty officer would respond to meet with the Fire Department Incident Commander. Once at the scene, the Coast Guard duty officer would confer with the Fire Department Incident Commander, then the Coast Guard would assume command of the entire operation (any incident in the water is the Coast Guard's jurisdiction). The Fire Department Incident Commander then becomes the Land Operations Section Chief.

Overall Incident Command would continue under the direction of the Coast Guard because an off-shore incident would be under Coast Guard jurisdiction. Two separate operations sections would be established under the Incident Commander: Land Operations Section and Ocean Operations Section. The Land Operations Section is managed by the Fire Department and is primarily responsible for patient triage, treatment, tracking, transportation, and security of the land area. The Ocean Operations Section is managed by the Coast Guard, and it is responsible for victim removal, extrication, triage and treatment at sea, and air operations, including helicopter coordination.

All victim extrication and removal is performed by the Coast Guard, unless assistance can be provided by helicopters and boats from outside agencies or military branches. Victims are transported from the ocean site to land by boat or helicopter. The helicopters deliver patients either to an ambulance on shore or directly to a hospital. Free-floating victims would be retrieved before victims in rafts. A boat shuttle system may be necessary to remove all victims. Initial triage and treatment could begin in the boats, especially if rescuers from shore eventually are assigned to the boats.

The Ocean Operations Section Commander establishes branches as necessary. Two important Ocean Operations branches are Rescue and Air Operations. The Rescue Branch Officer manages extrication of victims by boat.

The Air Operations Branch officer manages helicopter rescues and the landing zones, and any helicopter transport from land to hospitals.

Once on land, the patients would be turned over to the fire department, and ambulance personnel would function under the Land Operations Section. Three branches would be created under the Land Operations Section: Fire, Medical, and Police. The Fire Branch would provide firefighters to support either fire suppression or emergency medical needs. The Medical Branch would manage triage, treatment, tracking, and transport of victims. The Police Branch would manage scene security, traffic, and marine patrol. Victims would be moved through triage and treatment, and then transported. If necessary, decontamination occurs before transportation to the hospital. Treatment and transportation would be based on the triage priority (red tag, yellow tag, green tag). Helicopters and ambulances would transport the high-priority patients, while buses would transport the low-priority and ambulatory patients.

Additional support would be provided by the Planning, Finance, and Logistics Sections if necessary. The Planning Section is in charge of resource status, documentation, and technical specialties. The

Finance Section is responsible for procuring any special resources and tracking costs. The Logistics Section is in charge of special communications needs, food for rescuers, and managing special supplies at the scene.

There are two designated areas to establish Land Operations. Both areas are adjacent to docks where Coast Guard boats could off-load victims. The areas are large enough to establish a command post, a triage and treatment area, helicopter landing zones, and a staging area.

There are several reasons for designating two sites. Commercial ships in port may block access to the primary site. The secondary site is more protected from the ocean, allowing it to be used during rough sea periods.

Incident Command Position Responsibilities--The plan describes the duties of the major positions with the sections and branches of the Incident Command System. Pre-established position assignments and responsibilities eliminate confusion that could otherwise develop as multiple agencies work together on the same incident. The plan seeks to avoid a situation where two commanders have duplicate or overlapping responsibilities. It also allows the agencies to agree in advance about who is responsible for what so that multiple agencies are not each working independently to resolve the same problem. The primary incident command position responsibilities are listed below, along with the rationale for each position.

The *Incident Commander* (IC) is from the U.S. Coast Guard and directs all land and ocean rescue operations. Representatives from the different responding agencies report to the IC for direction and assignments. The IC has overall responsibility for the entire incident and establishes different components of the ICS structure, depending on the situation.

The *Ocean Operations Section Chief* a U.S. Coast Guard position, is in charge of coordinating all offshore rescue operations, including victim extrication and rescue and air rescue operations. The Ocean Operations Section Chief may direct operations from a Coast Guard boat or from land. This Section Chief determines what resources are necessary to effect the rescue, including boats and helicopters.

The *Land Operations Section Chief* is a Broward County Fire Department officer (Assistant Chief) responsible for coordinating the Fire, Police, Medical Branches, and Staging. Each agency working under the Land Operations Section Chief provides a representative as a liaison at the Command Post. This Section Chief determines what level of police, fire, and emergency medical response is necessary.

The *Liaison Officer* is the designated point of contact for all agency representatives reporting to the Command Post. The Liaison Officer directs agency representatives to the appropriate officer for assignment and acts as a filter to prevent congestion and keep unauthorized representatives from entering the Command Post. In order to establish a coordinated agency response from the outset, it is vital that this position be filled as soon as possible after the ICS is implemented.

One or more *Safety Officers* are assigned by the IC to monitor the safety of land, ocean, or air operations. There may be more than one safety officer, perhaps one from the Coast Guard monitoring operations at sea, and one from the fire department monitoring operations on land.

The *Public Information Officer* (PIO) is assigned by the IC to handle media personnel on scene and requests for information. PIOs from the other responding agencies report directly to the PIO assigned by the IC. The PIO releases information to the media after approval from the IC.

The *Land Staging Area Officer* is a fire department officer assigned by the Land Operation Section Chief to assist with the coordination of incoming resources. This officer controls the release of units from Staging to the scene as directed by the Land Operations Section Chief

The person from each responding agency who is assigned to the Command Post maintains contact through the Liaison Officer. This allows the Incident Commander to coordinate with the other agencies and facilitates efficient communications among the responding agencies. A restricted number of representatives are allowed into the Command Post to avoid confusion. Each of the following agencies is represented by one person at the command post:

Airport (management)
Airport Public Information Office
Airport Fire and Rescue Department
Broward County Fire Rescue
Broward EMS
Broward County Emergency Management
Port Everglades Fire and Safety
Broward Sheriff's Office
Florida Marine Patrol
Other local fire agencies
Other local law enforcement agencies
Safety Officer
Liaison

The Land Operations Section includes the following individuals:

Land Operations Section Chief (Broward County Assistant Chief)
Fire Branch Officer (Broward County Fire and Rescue)
Police Branch Officer (Broward Sheriff's Office)
Medical Branch Officer (Broward County Fire and Rescue)
Staging Officer (first arriving Port Everglades Officer)

The Ocean Operations Section comprises the following personnel:

Operation Section Chief (U.S. Coast Guard Boat Officer)
Rescue Branch (U.S. Coast Guard Officer)
Air Operation Branch (U.S. Coast Guard Officer)
Ocean Staging (U.S. Coast Guard Officer)

Communications--One of the most important aspects of managing any large-scale incident is effective communications. Without this, the organization and control of the incident are severely compromised because information and instructions cannot be communicated in a timely manner.

Any rescue plan that combines resources from multiple agencies must consider how to effectively integrate communications between the agencies. The radio communications plan involves four primary channels:

U.S. Coast Guard Operations Channel (VHF)
Air Operations Channel (VHF)
Fire Operations Channel (800 Mhz)
Medical Operations Channel (800 MHz)

The Ocean Rescue Mass Casualty Plan is designed to coordinate communications between agencies, but the communication patterns are complex because most of the agencies operate on separate and incompatible radio systems. No frequencies are patched together in the plan. The link between agencies occurs at the main command post and at section command posts. The primary point of contact is the command post, where messages are relayed between agency representatives and field units.

DESIGNING THE DISASTER PLAN AND DRILL

The goal of the planners of the mass casualty drill was to conduct a realistic and safe drill that would test the disaster response plan and train personnel from the responding agencies about response to a major disaster.

Coordination of operations and logistics was the greatest challenge because the agencies had never worked together before as a group. The plan was designed primarily by officials from the Hollywood/ Fort Lauderdale International Airport and the Broward County Fire Department. The airport and fire department officials met with officials from each of the participating emergency agencies to get their input and approval of the plan. In the weeks before the drill, each agency was given an overview of the disaster plan and an opportunity to practice their role in the plan. Tabletop scenarios were used in some cases. Some of the special considerations that were made by the planning group to ensure that the drill would be realistic and safe are discussed below.

Notification of Drill--When planning a drill, it is important to notify any public or private entities which may be affected by the drill. All of the agencies participating in the drill were informed by virtue of the fact that they had representatives who had approved their agencies' roles in the drill. Some of the entities affected included: shipping companies, the Port Authority, the Federal Aviation Administration and the airport control tower; as well as neighboring agencies not participating in the drill; and the news media.

Evaluators--A team of evaluators was formed to critique drill operations. The team was comprised of representatives from various groups that would be affected by a disaster and included: doctors, paramedics, firefighters, police, and pilots. Each evaluator was given a form to critique different aspects of the response. Some of the evaluators were assigned to examine specific areas of the response such as incident command, victim treatment, and transport. Others monitored the response as a whole.

Victims--Neighboring schoolchildren, friends, and family members of rescuers, and flight attendants served as victims for the drill. Over 200 victims were necessary to realistically simulate the number of patients posed by a downed commercial aircraft. The victims were instructed to meet at a central meeting point several hours before the drill. They were briefed about what would happen and then each was given a card describing their injuries. Moulage was applied to many of the victims. Approximately 50 victims were selected to be rescued from the water. These victims had to pass a swimming test and were given special instructions before they were allowed participate.

Drill Overview for Rescuers--Tabletop and practice sessions were necessary to provide an overview of the plan before it was tested in the drill. Each participating emergency agency was given an overview of the plan and the drill, which included training on the incident command system. Aerial view photographs were used to show rescuers the layout of the disaster scene and the two predetermined sites for locating the Land Operations Section.

Weather--Although many aspects of the drill could be planned for, the weather was an exception. Because the drill involved victims and rescuers in the water, boats, and helicopters, the drill planners set parameters in advance for acceptable weather conditions. If the weather conditions were too severe, the drill would be postponed for safety purposes.

Critique Session--A critique session was scheduled several days after the drill to allow time for participants to rest and for organizers to gather notes from the evaluators. The critique session was an important component of the drill to reveal weaknesses in the response plan and lessons learned.

THE DRILL

The drill began shortly after 10:00 a.m. on December 6, 1994. Two sections of mock aircraft floating fuselage were placed in the water to simulate the crash, but the 50 in-water victims were free of the wreckage and extrication was not necessary (see diagram in Appendix C). The remaining victims were bused-in to simulate victims swimming to shore or transported to shore by boat.

The first arriving fire units found victims floating in the water, and shortly thereafter, additional victims arriving on land. As planned, the fire department took command until the Coast Guard commanding officer arrived. The Coast Guard then assumed command of the overall incident with a command center on land. The fire department commander became the Land Operations officer.

Coast Guard helicopters and boats quickly arrived on the scene to begin the rescue of victims from the water. For the purposes of the drill, fire department personnel did not rescue victims from the water even though victims were only 50 feet off-shore, because in real circumstances the victims could be miles off-shore. Additional fire units responded, including some from the airport.

Off-shore, the Coast Guard established rescue and air operations sectors to manage victim retrieval. All in-water victims were removed by boat. The air operations section handled incoming helicopters which were used to transport the most critical patients from shore to trauma centers. All in-water victims were removed within fifteen minutes of the start of the drill.

Meanwhile, on-shore, the fire department established triage, treatment, and transportation sectors under the Land Operations Section. Patients were quickly moved through triage and treatment; however, a bottleneck situation occurred in the transportation area due to an inadequate number of ambulances. Rescuers found themselves inundated by the number of victims, which turned out to be nearly 250. Because only a limited number of ambulances could be committed to the drill, some "red tag" and "yellow tag" patients had to be transported by bus with the "green tag" patients.

The Broward County hazardous materials team set up a decontamination area to clean patients who had been exposed to jet fuel. A morgue area was set up for fatally injured victims.

The drill terminated around 1 p.m. after the final victims were transported, nearly three hours after the mock crash occurred.

LESSONS LEARNED AND REINFORCED FROM THE DRILL

Many lessons were learned from this drill. Some of the lessons are specific to an ocean rescue disaster scenario, while others apply to virtually any disaster scenario. Below is a list of some of the lessons learned or reinforced from this drill.

1. **It is important to test a disaster plan by conducting an organized drill.** The drill will expose strengths and weaknesses in the plan and in the capabilities of responders. It also allows various agencies that do not commonly work together to become familiar with one another's response capabilities and with their roles in the plan.

2. **A command board with a master checklist of tasks to be performed helps the Incident Commander stay organized while managing the incident.** It is also necessary to keep track of personnel and resources.

3. **Radio traffic should be limited to transmissions between commanders and sector officers.** Other responders should avoid making transmissions unless there is a need to convey emergency information.

4. **Personnel (possibly an engine company' may have to be assigned as runners to improve communications.**

5. **The Staging Officer should establish one entry/exit point to the scene if possible.** A fire department officer should be posted at this point to prevent unauthorized entry by any unit and to avoid freelancing by incoming units. Police officers should not be used for this function because they would not know which units have been authorized for entry by the fire department Incident Commander.

6. **The Liaison Officer's job is critical to maintain order at the command post.** The Liaison Officer should set up operations near the post and control access to the Command Post.

7. **The Ground Operations is area must be large enough to handle multiply vehicles, patients, helicopters, the media, and sector areas (triage, treatment, etc.).** The area used for this drill was too small and created confusion. (The larger areas designated in the response plan could not be used in this drill.)

8. **Air Operations control is vitally important when multiple helicopters are called to the scene.** Control can be provided by a nearby airport tower or even by one designated hovering helicopter. All helicopters must be equipped to talk to the air controller and to the hospitals.

9. **Helicopter landing zones should be kept distant from the incident command, triage, and treatment areas.** The noise and flying debris generated by helicopters can severely disrupt operations in these areas.

10. **It is imperative that all early arriving units strategically position their apparatus to leave entry and exit lanes and parking space for other incoming units.**

11. **A public information officer should be available at drills to control the media.**

12. **Outside agencies who are not aware that a drill is taking place may respond to the scene without being requested under the belief that a real incident is taking place.**

13. **Records of patient transports are important to provide to officials handling inquiries from relatives of those injured.**

14. **The priority of the first incoming EMS unit is to determine the magnitude of the incident and estimate the number of victims but not to assist the first critical victim found.** Organized EMS operations from the outset will allow EMS commanders to determine the number of ambulances and helicopters that will be needed.

15. **All area hospitals should be contacted as soon as the magnitude of the incident is known.** This will allow hospitals to gear up for the incoming patients and to make their own personnel call backs. Rescuers should determine the number of red, yellow, and green tag patients that each hospital can accept so that no one hospital is overwhelmed. Green tag patients should be sent to distant hospitals, whereas yellow and red tags should be sent to the nearest trauma hospitals.

16. **Commanders should wear vests designating their command functions.** Sector areas should be designated by signs, which will help guide victims unfamiliar with emergency operations through the triage, treatment, and transport system.

17. **It is advantageous to carry spare portable radios and batteries in the command vehicle.** These radios can be distributed to incoming units which do not have access to the radio channel for the scene.

18. **The presence of jet fuel creates additional hazards for victims and rescuers.** If at all possible, utilize Hazmat teams to decontaminate victims at the scene because hospitals may not be equipped to handle a contaminated patient.

APPENDIX A

Broward County Ocean Rescue Mass Casualty Plan

1.0   PURPOSE:

The United States Coast Guard and Broward County have agreed upon a unified incident command system to handle large-scale ocean rescues.

1.1   The intent of this plan is to provide an aircraft ocean rescue mass casualty incident command system at Port Everglades and to assign functional responsibilities to responding agencies.

1.2   COMMAND:

1.2.1   The Initial establishment of command will be done by Broward County Fire Rescue Battalion 6 or a Port Everglades fire officer. The United States Coast Guard (USCG) Incident Commander, once alerted, will respond from the Port Everglades station, to the assigned command post and assume command. Battalion 6 will then assume the position of Land Operation Section Chief until relieved. Battalion 6 will maintain contact with the USCG on the assigned marine radio channel.

1.2.2   A **primary** command post, staging area, medical triage area, helispots, and land operations section will be located on the west side of the 17th Street Causeway bridge in the parking lot behind the Broward Convention Center and the Sea Escape Cruise Line.

1.2.3   A **secondary** site will be located on the Port Everglades Clean-Up Committee property 3510 SE 19th Avenue, Port Everglades, FL. This site will be used when the turn basin water is extremely rough and will not permit boats to dock and remove victims safely.

1.2.4   All command post personnel listed in the plan will report to the Liaison Officer at the land command post.

1.3   RESPONSIBILITIES:

1.3.1   **Incident Commander (IC)**: The Incident Commander (USCG) will direct all land and ocean rescue operations. Various agency representatives listed in the plan will report to the command post to assist the IC and to deal directly with their agency and/or area of responsibility at the request of Command.

1.3.2   **Ocean Operation Section Chief**: The USCG will be responsible for all offshore rescue operations. The boat officer will maintain contact and coordinate with the IC and the Land Operation Section Chief. In addition, this position will be responsible for air rescue operations.

1.3.3   **Land Operation Section Chief**: The Land Operation Section Chief will be responsible for coordinating the fire, police, medical branches, and staging area. Designated agencies will provide a representative to work with the Operation Chief. The Operation Chief will determine the required personnel and equipment needs for the land operation. All unauthorized units responding or arriving at the scene will be returned to their place of origin.

1.3.4   **Liaison Officer**: The Liaison Officer will be assigned by Command. The officer will be the point of contact for all agency representatives reporting to the Command Post. This position needs to be established by Command as soon as possible.

1.3.5   **Fire Branch**: The Fire Branch will provide personnel and equipment to land or ocean operational needs as requested by the Operation Section Chief.

1.3.6   **Police Branch**: The Police Branch will provide security for land and ocean operations. In addition, they will provide traffic control for transport of victims to various hospitals. BSO will be designated as branch officer under the direction of the Operation Section Chief. BSO will coordinate city, county, and state patrol boats.

1.3.7   **Medical Branch**: The Medical Branch will establish a triage area to assess victims brought to shore. Command may request personnel be sent to various USCG boats to provide triage and treatment at sea. Broward County Fire Rescue will be designated as the medical branch.

1.3.8   **Ocean Rescue Branch**: USCG will coordinate all victim removal and transport to shore.

1.3.9   **Air Operations Branch**: USCG will coordinate all air operations (police, news media, etc.) in the designated land and ocean rescue area. A clear area will be established for all aircraft not assigned to rescue operations.

1.3.10 **Safety Officer:** The Safety Officer will be assigned by command. Qualified personnel from the fire branch may be used for land or ocean safety observers.

1.3.11 **Information Officer (PIO)**: A lead PIO will be assigned by Command. All authorized PIO's will coordinate rescue information with the lead PIO before release to the media. No information will be released without approval of Command.

1.3.12 **Planning Section**: The Planning Section will be established by Command.

1.3.13 **Logistics Section**: The Logistics Section will be established by Command.

1.3.14 **Finance Section**: The Finance Section will be established by Command. All agencies will be responsible for their own personnel and equipment costs.

1.3.15 **Staging Area Officer**: The Staging Area Officer will be assigned by the Land Operation Section Chief. The Staging Officer will coordinate all resources with the Land Operation Section Chief. Police road blocks may be needed to stop unauthorized mutual aid from entering the Port Everglades area. NO units will be permitted to leave the staging area without approval of the Staging Officer.

1.4   COMMAND POST PERSONNEL: ONE PERSON PER AGENCY IN THE CP

INCIDENT COMMANDER--USCG
AIRPORT AGENT
AIRPORT PIO
AIRPORT ARFF
BROWARD COUNTY FIRE RESCUE
BROWARD COUNTY EMERGENCY MANAGEMENT
PORT EVERGLADES FIRE
BROWARD SHERIFF OFFICE
FLA. MARINE PATROL
LOCAL FIRE AGENCY
LOCAL POLICE AGENCY
SAFETY OFFICER
LIAISON

1.5   LAND OPERATION SECTION PERSONNEL:

Operation Section Chief--Broward County Assistant Chief
Fire Branch Officer--Broward County Fire Rescue
Battalion Chief
Police Branch Officer--BSO
Medical Branch Officer--Broward County Fire Rescue
Division Chief
Staging Officer--First arrival Port Everglades Engine

1.6   OCEAN OPERATION SECTION PERSONNEL:

Operation Section Chief--USCG boat officer
Rescue Branch--USCG assigned officer
Air Operation Branch--USCG assigned officer
Ocean Staging--USCG assigned officer

1.7   COMMUNICATIONS:

Incident Command to Land and Ocean Operations Sections
ch. 23 (Marine radio)

Ocean Operation Section to air operations
ch. 21 (Marine band)

Airport Tower and helicopter communications
ch. 123.05 VHF (Aircraft band)

Land Operation Section to IC and USCG boat
ch. 23 (Marine band) (Verbal communications with fire, police and medical branch officers).

Medical Branch to triage area and boat medics
ch. 141 (800mhz)

Fire Branch to staging area and division/groups
ch. 14J (800mhz)

Police Branch to Marine Patrol units, traffic control
system 21RP (800mhz)

Ocean Operations Section to civilian craft in rescue area.
ch. 81 (Marine band)

1.8   Assigned command post personnel will contact their respective units on their own radio channels. Verbal coordination will be used within the command post.

APPENDIX B

Florida Statewide Mutual Aid Plan

April 27, 1994

STATEWIDE MUTUAL AID AGREEMENT FOR CATASTROPHIC DISASTER RESPONSE AND RECOVERY

THIS AGREEMENT IS ENTERED INTO BETWEEN THE STATE OF FLORIDA, DIVISION OF EMERGENCY MANAGEMENT AND AMONG EACH POLITICAL SUBDIVISION OF THE STATE THAT EXECUTES AND ADOPTS THE TERMS AND CONDITIONS CONTAINED HEREIN, BASED UPON THE FOLLOWING FACTS:

WHEREAS, the State Emergency Management Act, Chapter 252, Florida Statutes, authorizes the state and its political subdivisions to develop and enter into mutual aid agreements for reciprocal emergency aid and assistance in case of emergencies too extensive to be dealt with unassisted; and

WHEREAS, Chapter 252, Florida Statutes, sets forth details concerning powers, duties, rights, privileges, and immunities of political subdivisions of the state rendering outside aid; and

WHEREAS, Chapter 252, Florida Statutes, authorizes the State to enter into a contract on behalf of the state for the lease or loan to any political subdivision of the state any real or personal property of the state government or the temporary transfer or employment of personnel or the state government to or by any political subdivision of the state; and

WHEREAS, Chapter 252, Florida Statutes, authorizes the governing body of each political subdivision of the state to enter into such contract or lease with the state, accept any such loan, or employ such personnel, and such political subdivision may equip, maintain, utilize, and operate any such property and employ necessary personnel therefor in accordance with the purposes for which such contract is executed, and to otherwise do all things and perform any and all acts which it may deem necessary to effectuate the purpose for which such contract was entered into; and

WHEREAS, Chapter 252, Florida Statutes, authorizes the Division of Emergency Management to make available any equipment, services, or facilities owned or organized by the state or its political subdivisions for use in the affected area upon request of the duly constituted authority of the area or upon the request of any recognized and accredited relief agency through such duly constituted authority; and

WHEREAS, Chapter 252, Florida Statutes, authorizes the Division of Emergency Management to call to duty and otherwise provide, within or without the state, such support from available personnel, equipment, and other resources of state agencies and the political subdivisions of the state as may be necessary to reinforce emergency management agencies in areas stricken by emergencies; and

WHEREAS, Chapter 252, Florida Statutes, requires that each municipality must coordinate requests for state or federal emergency response assistance with its county; and

WHEREAS, the State of Florida is geographically vulnerable to hurricanes, tornadoes, freshwater flooding, sinkhole formations, and other natural disasters that in the past have caused severe disruption of essential human services and severe property damage to public roads, utilities, buildings, parks, and other government owned facilities; and

WHEREAS, the Parties to this Agreement recognize that additional manpower and equipment may be needed to mitigate further damage and restore vital services to the citizens of the affected community should such disasters occur; and

WHEREAS, to provide the moss effective mutual aid possible, each Participating Government, intends to foster communications between the personnel of the other Participating Government by visits, compilation of asset inventories, exchange of information and development of plans and procedures to implement this Agreement;

NOW, THEREFORE, the Parties hereto agree as follows:

SECTION 1. DEFINITIONS

A. "AGREEMENT"--the Statewide Mutual Aid Agreement for Emergency Response/Recovery. Political subdivisions of the State of Florida may become a party to this Agreement by executing a copy of this Agreement and providing a copy with original signatures and authorizing resolution(s) to the State of Florida Division of Emergency Management. Copies of the agreement with original signatures and copies of authorizing resolutions and insurance letters shall be filed and maintained at the Division headquarters in Tallahassee, Florida.

B. "REQUESTING PARTY"--the participating government entity requesting aid in the event of an emergency. Each municipality must coordinate requests for state or federal emergency response assistance through its county.

C "ASSISTING PARTY"--the participating government entity furnishing equipment, services and/ or manpower to the Requesting Party.

D "AUTHORIZED REPRESENTATIVE"--an employee of a participating government authorized in writing by that government to request, offer, or provide assistance under the terms of this Agreement. The list of authorized representatives for the participating government executing this Agreement shall be attached as Appendix A to the executed copy of the Agreement supplied to the Division, and shall be updated as needed by each participating government.

E. "DIVISION"--the State of Florida, Department of Community Affairs, Division of Emergency Management.

F. "EMERGENCY"--any occurrence, or threat thereof, whether natural, or caused by man, in war or in peace, which results or may result in substantial injury or harm to the population or substantial damage to or loss of property.

G. "DISASTER"--any natural, technological, or civil emergency that causes damage of sufficient severity and magnitude to result in a declaration of a state of emergency by a county, Governor, or the President of the United States.

H. "PARTICIPATING GOVERNMENT"--the State of Florida and any political subdivision of the State of Florida which executes this mutual aid agreement and supplies a complete executed copy to the Division.

I. "PERIOD OF ASSISTANCE"--the period of time beginning with the departure of any personnel of the Assisting Party from any point for the purpose of traveling to the Requesting Party in order to provide assistance and ending upon the return of all personnel and equipment of the Assisting Party, after providing the assistance requested, to their residence or regular place of work, whichever occurs first. The period of assistance shall not include any portion of the trip to the Requesting Party or the return trip from the Requesting Party during which the personnel of the Assisting Party are engaged in a course of conduct not reasonably necessary for their safe arrival at or return from the Requesting Party.

J. "WORK OR WORK-RELATED PERIOD"--any period of time in which either the personnel or equipment of the Assisting Party are being used by the Requesting Party to provide assistance and for which the Requesting Party will reimburse the Assisting Party. Specifically included within such period of time are rest breaks when the personnel of the Assisting Party will return to active work within a reasonable time. Specifically excluded from such period of time are breakfast, lunch, and dinner breaks.

SECTION 2. PROCEDURES

When a participating government either becomes affected by, or is under imminent threat of, an emergency or disaster, it may invoke emergency related mutual aid assistance either by: (i) declaring a state of local emergency and transmitting a copy of that declaration to the Assisting Party, or to the Division, or (ii) by orally communicating a request for mutual aid assistance to Assisting Party or to the Division, followed as soon as practicable by written confirmation of said request. Mutual aid shall not be requested by Participating Governments unless resources available within the stricken area are deemed inadequate by the Local Emergency Management Agency. All requests for mutual aid shall be transmitted by the Authorized Representative or the Director of the Local Emergency Management Agency. Requests for assistance may be communicated either to the Division or directly to an Assisting Party. Requests for assistance under this Agreement shall be limited to catastrophic disasters, except where the Participating Government has no other mutual aid agreement based upon Section 252.40 or 163.01, Florida Statutes, in which case a Participating Government may request assistance pursuant to the provisions of this agreement.

A. REQUESTS DIRECTLY TO ASSISTING PARTY: The Requesting Party may directly contact the authorized representative of the Assisting Party and shall provide them with the information in paragraph C below. All communications shall be conducted directly between the Requesting and Assisting Party. Each party shall be responsible for keeping the Division advised of the status of the response activities. The Division shall not be responsible for costs associated with such direct requests for assistance. However, the Division may provide, by rule, for reimbursement of eligible expenses from the Emergency Management Preparedness and Assistance Trust Fund created under Section 252.373, Florida Statutes.

B. REQUESTS ROUTED THROUGH, OR ORIGINATING FROM, THE DIVISION: The Requesting Party may directly contact the Division, in which case it shall provide the Division with the information in paragraph C below. The Division may then contact other Participating Governments on behalf of the Requesting Party and coordinate the provision of mutual aid. The Division shall not be responsible for costs associated with such indirect requests for assistance, unless the Division so indicates in writing at the time it transmits the request to the Assisting Party. In no event shall the Division or the State of Florida be responsible for costs associated with assistance in the absence of appropriated funds. In all cases, the party receiving the mutual aid shall be primarily responsible for the costs incurred by any Assisting Party providing assistance pursuant to the provisions of this Agreement.

C. REQUIRED INFORMATION: Each request for assistance shall be accompanied by the following information, to the extent known:

1. A general description of the damage sustained;

2. Identification of the emergency service function for which assistance is needed (e.g., fire, law enforcement, emergency medical, transportation, communications, public works and engineering, building, inspection, planning and information assistance, mass care, resource support, health and other medical services, search and rescue, etc.) and the particular type of assistance needed;

3. Identification of the public infrastructure system for which assistance is needed (e.g., sanitary sewer, potable water, streets, or storm water systems) and the type of work assistance needed;

4. The amount and type of personnel, equipment, materials, and supplies needed and a reasonable estimate of the length of time they will be needed;

5. The need for sites, structures or buildings outside the Requesting Party's political subdivision to serve as relief centers or staging areas for incoming emergency goods and services; and

6. A specific time and place for a representative of the Requesting Party to meet the personnel and equipment of any Assisting Party.

This information may be provided on the form attached as Exhibit B, or by any other available means. The Division may revise the format of Exhibit B subsequent to the execution of this agreement, in which case it shall distribute copies to all participating governments.

D. ASSESSMENT OF AVAILABILITY OF RESOURCES AND ABILITY TO RENDER ASSISTANCE: When contacted by a Requesting Party or the Division the authorized representatives of any participating government agree to assess their government's situation to determine available personnel, equipment and other resources. All participating governments shall render assistance to the extent personnel, equipment and resources are available. Each participating government agrees to render assistance in accordance with the terms of this Agreement to the fullest extent possible. When the authorized representative determines that his Participating Government has available personnel, equipment or other resources, they shall so notify the Requesting Party or the Division, whichever communicated the request, and provide the information below. The Division shall, upon response from sufficient participating parties to meet the needs of the Requesting Party, notify the authorized representative of the Requesting Party and provide them with the following information, to the extent known:

1. A complete description of the personnel, equipment, and materials to be furnished to the Requesting Party;

2. The estimated length of time the personnel, equipment, and materials will be available;

3. The areas of experience and abilities of the personnel and the capability of the equipment to be furnished;

4. The name of the person or persons to be designated as supervisory personnel; and

5. The estimated time when the assistance provided will arrive at the location designated by the authorized representative of the Requesting Party.

E. SUPERVISION AND CONTROL: The personnel, equipment and resources of any Assisting Party shall remain under operational control of the Requesting Party for the area in which they are serving. Direct supervision and control of said personnel, equipment and resources shall remain with the designated supervisory personnel of the Assisting Party. Representatives of the Requesting Party shall provide work tasks to the supervisory personnel of the Assisting Party. The designated supervisory personnel of the Assisting Party shall have the responsibility and authority for assigning work and establishing work schedules for the personnel of the Assisting Party, based on task or mission assignments provided by the Requesting Party and the Division. The designated supervisory personnel of the Assisting Party shall: maintain daily personnel time records, material records and a log of equipment hours; be responsible for the operation and maintenance of the equipment and other resources furnished by the Assisting Party; and shall report work progress to the Requesting Party. The Assisting Party's personnel and other resources shall remain subject to recall by the Assisting Party at any time, subject to reasonable notice to the Requesting Party and the Division. At least twenty-four hour advance notification of intent to withdraw personnel or resources shall be provided to the Requesting Party unless such notice is not practicable, in which case such notice as is reasonable shall be provided.

F. FOOD; HOUSING; SELF-SUFFICIENCY: Unless specifically instructed otherwise, the Requesting Party shall have the responsibility of providing food and housing for the personnel of the Assisting Party from the time of their arrival at the designated location to the time of their departure. However, Assisting Party personnel and equipment should be, to the greatest extent possible, self-sufficient for operations in areas stricken by emergencies or disasters. The Requesting Party may specify only self-sufficient personnel and resources in its request for assistance.

G. COMMUNICATIONS: Unless specifically instructed otherwise, the Requesting Party shall have the responsibility for coordinating communications between the personnel of the Assisting Party and the Requesting Party. Assisting Party personnel should be prepared to furnish communications equipment sufficient to maintain communications among their respective operating units.

H. RIGHTS AND PRIVILEGES: Whenever the employees of the Assisting Party are rendering outside aid pursuant to this Agreement, such employees shall have the powers, duties, rights, privileges, and immunities, and shall receive the compensation, incidental to their employment.

I. WRITTEN ACKNOWLEDGEMENT: The Requesting Party shall complete a written acknowledgment regarding the assistance to be rendered, setting forth the information transmitted in the request, and shall transmit it by the quickest practical means to the Assisting Party or the Division, as applicable, for approval. The form to serve as this written acknowledgement is attached as Attachment C. The Assisting Party/Division shall respond to the written acknowledgement by executing and returning a copy to the Requesting Party by the quickest practical means, maintaining a copy for its files.

SECTION 3. REIMBURSABLE EXPENSES

The terms and conditions governing reimbursement for any assistance provided under this Agreement shall be in accordance with the following provisions, unless otherwise agreed upon by the Requesting and Assisting Parties and specified in the written acknowledgment executed in accordance with paragraph 2.I of this Agreement. The Requesting Party shall be ultimately responsible for reimbursement of all reimbursable expenses.

A. PERSONNEL--During the period of assistance, the Assisting Party shall continue to pay its employees according to its then prevailing ordinances, rules, and regulations. The Requesting Party shall reimburse the Assisting Party for all direct and indirect payroll costs and expenses incurred during the period of assistance, including, but not limited to, employee pensions and benefits as provided by Generally Accepted Accounting Principles (GAAP). The Requesting Party shall reimburse any amounts paid or due for compensation to employees of the Assisting Party under the terms of the Florida Workers' Compensation Act due to personal injury or death occurring while such employees are engaged in rendering aid under this Agreement. While providing services to the Requesting Party, employees of the Assisting Party shall be considered "borrow servants" of the Requesting Party and shall be considered in the "dual employment" with the Requesting and Assisting Parties, subject to the supervision and control of both for purposes of Chapter 440, Florida Statutes. While the Requesting Party shall reimburse the Assisting Party for payments made in workers' compensation benefits required to be paid to its employees due to personal injury or death, the Division, and both the Requesting and Assisting Party shall enjoy immunity from civil prosecution as provided for in the Florida Workers' Compensation Act.

B. EQUIPMENT--The Assisting Party shall be reimbursed by the Requesting Party for the use of its equipment during the period of assistance according to either a pre-established local or state hourly rate or according to the actual replacement, operation, and maintenance expenses incurred. For those instances in which costs are reimbursed by the Federal Emergency Management Agency, the eligible direct costs shall be determined in accordance with 44 CFR 206.228. The Assisting Party shall pay for all repairs to its equipment as determined necessary by its on-site supervisor(s) to maintain such equipment in safe and operational condition. At the request of the Assisting Party, fuels, miscellaneous supplies, and minor repairs may be provided by the Requesting Party, if practical. The total equipment charges to the Requesting Party shall be reduced by the total value of the fuels, supplies, and repairs furnished by the Requesting Party and by the amount of any insurance proceeds received by the Assisting Party.

C. MATERIALS AND SUPPLIES--The Assisting Party shall be reimbursed for all materials and supplies furnished by it and used or damaged during the period of assistance, except for the costs of equipment, fuel and maintenance materials, labor and supplies, which shall be included in the equipment rate established in 3.B. above, unless such damage is caused by gross negligence, willful and wanton misconduct, intentional misuse, or recklessness of the Assisting Party's personnel. The Assisting Party's Personnel shall use reasonable care under the circumstances in the operation and control of all materials and supplies used by them during the period of assistance. The measure of reimbursement shall be determined in accordance with 44 CFR 206.228. In the alternative, the Parties may agree that the Requesting Party will replace, with like kind and quality as determined by the Assisting Party, the materials and supplies used or damaged. If such an agreement is made, it shall be reduced to writing and transmitted to the Division.

D. RECORD KEEPING--The Assisting Party shall maintain records and submit invoices for reimbursement by the Requesting Party or the Division using format used or required by FEMA publications, including 44 CFR part 13 and applicable Office of Management and Budget Circulars. Requesting Party and Division finance personnel shall provide information, directions, and assistance for record keeping to Assisting Party personnel.

E. PAYMENT--Unless otherwise mutually agreed in the written acknowledgement executed in accordance with paragraph 2.I. or a subsequent written addendum to the acknowledgement, the Assisting Party shall bill the Requesting Party for all reimbursable expenses with an itemized Notice as soon as practicable after the expenses are incurred, but not later than sixty (60) days following the period of assistance, unless the deadline for identifying damage is extended in accordance with 44 CFR part 206. The Requesting Party shall pay the bill, or advise of any disputed items, not later than sixty (60) days following the billing date. These timeframes may be modified by mutual agreement. This shall not preclude an Assisting Party or Requesting Party from assuming or donating, in whole or in part, the costs associated with any loss, damage, expense or use of personnel, equipment and resources provided to a Requesting Party.

F. PAYMENT BY OR THROUGH THE DIVISION--The Division of Emergency Management may reimburse for all actual and necessary travel and subsistence expenses for personnel providing assistance pursuant to the request of the Division, to the extent of funds available, and contingent upon an annual appropriation from the Legislature for such purposes. The Assisting Party shall be responsible for making written requests to the Division for reimbursement of travel and subsistence expenses, prior to submitting a request for payment to the Requesting Party. The Assisting Party's written request should be submitted as soon as possible after expiration of the period of assistance. The Division shall provide a written response to said requests within ten (10) days of actual receipt. If the Division denies said request, the Assisting Party shall then bill the Requesting Party. In the event that an affected jurisdiction requests assistance without forwarding said request through the Division, or an assisting party provides assistance without having been requested by the Division to do so, the Division shall not be liable for reimbursement of any of the cost(s) of assistance. The Division may serve as the eligible entity for requesting reimbursement of eligible costs from FEMA. Any costs to be so reimbursed by or through the Division shall be determined in accordance with 44 CFR 206.228. The Division may authorize applications for reimbursement of eligible costs from the undeclared disaster portion of the Emergency Management Preparedness and Assistance Trust Fund established pursuant to Section 252.373, Florida Statutes, in the event that the disaster or emergency event is not declared pursuant to the Robert T. Stafford Disaster Relief and Emergency Assistance Act, Public Law 93-288, as amended by Public Law 100-707. Such applications shall be evaluated pursuant to rules established by the Division, and may be funded only to the extent of available funds.

SECTION 4. INSURANCE

Each participating government shall bear the risk of its own actions, as it does with its day-to-day operations, and determine for itself what kinds of insurance, and in what amounts, it should carry. If a participating government is insured, its file shall contain a letter from its insurance carrier authorizing it to provide and receive assistance under this Agreement, and indicating that there will be no lapse in its insurance coverage either on employees, vehicles, or liability. If a participating government is self-insured, its file shall contain a copy of a resolution authorizing its self-insurance program. A copy of the insurance carrier's letter or the resolution of self-insurance shall be attached to the executed copy of this Agreement which is filed with the Division. Each Assisting Party shall be solely responsible for determining that its insurance is current and adequate prior to providing assistance under this agreement. The amount of reimbursement from the Division or the Requesting Party shall be reduced by the amount of any insurance proceeds to which the Assisting Party is entitled as a result of losses experienced in rendering assistance pursuant to this Agreement.

SECTION 5. LIABILITY

To the extent permitted by law, and without waiving sovereign immunity, each Party to this Agreement shall be responsible for any and all claims, demands, suits, actions, damages, and causes of action related to or arising out of or in any way connected with its own actions, and the actions of its personnel, in providing mutual aid assistance rendered or performed pursuant to the terms and conditions of this Agreement.

SECTION 6. LENGTH OF TIME FOR EMERGENCY

The duration of such state of emergency declared by the Requesting Party is limited to seven (7) days. It may be extended, if necessary, in 7 day increments.

SECTION 7. TERM

This Agreement shall be in effect for one (1) year from the date hereof and shall automatically be renewed in successive one (1) year terms unless terminated in writing by the participating government. Notice of such termination shall be made in writing and shall be served personally or by registered mail upon the Director, Division of Emergency Management, Florida Department of Community Affairs, Tallahassee, Florida, which shall provide copies to all other Participating Parties.

SECTION 8. EFFECTIVE DATE OF THIS AGREEMENT

This Agreement shall be in full force and effect upon approval by the participating government and upon proper execution hereof.

SECTION 9. ROLE OF DIVISION OF EMERGENCY MANAGEMENT

The responsibilities the Division of Emergency Management, Florida Department of Community Affairs under this Agreement are to: (1) request mutual aid on behalf of a participating government, under the circumstances identified in this Agreement; (2) coordinate the provision of mutual aid to a requesting party, pursuant to the provisions of this Agreement; (3) serve as the eligible entity for requesting reimbursement of eligible costs from FEMA, upon a Presidential disaster declaration; (4) serve as central depository for executed Agreements; and (5) maintain a current listing of Participating Governments with their Authorized Representative and contact information, and to provide a copy of the listing to each of the Participating Governments on an annual basis during the second quarter of the calendar year.

SECTION 10. SEVERABILITY; EFFECT ON OTHER AGREEMENTS

Should any portion, section, or subsection of this Agreement be held to be invalid by a court of competent jurisdiction, that fact shall not affect or invalidate any other portion, section or subsection; and the remaining portions of this Agreement shall remain in full force and affect without regard to the section, portion, or subsection or power invalidated.

In the event that any parties to this agreement have entered into other mutual aid agreements, pursuant to Section 252.40, Florida Statutes, or interlocal agreements, pursuant to Section 163.01, Florida Statutes, those parties agree that said agreements are superseded by this agreement only for emergency management assistance and activities performed in catastrophic emergencies pursuant to this agreement. In the event that two or more parties to this agreement have not entered into another mutual aid agreement, and the parties wish to engage in mutual aid, then the terms and conditions of this agreement shall apply unless otherwise agreed between those parties.

IN WITNESS WHEREOF, the parties set forth below have duly executed this Agreement on the date set forth below:

ATTEST: BOARD OF
CLERK OF THE CIRCUIT COURT   OF _____ FLORIDA
                                                        (COUNTY)

By: _____   By: _____
            DEPUTY CLERK   CHAIRMAN

                                           APPROVED AS TO FORM:
                                           Office of the Attorney

                                           By: _____

EXECUTED BY THE FOLLOWING PARTICIPATING LOCAL GOVERNMENTS IN _____
COUNTY (attach authorizing resolution or ordinance and insurance letter or resolution for each).

_____   by _____   _____
     POLITICAL SUBDIVISION   AUTHORIZED OFFICIAL   DATE

_____   by _____   _____
     POLITICAL SUBDIVISION   AUTHORIZED OFFICIAL   DATE

_____   by _____   _____
     POLITICAL SUBDIVISION   AUTHORIZED OFFICIAL   DATE

_____   by _____   _____
     POLITICAL SUBDIVISION   AUTHORIZED OFFICIAL   DATE

_____   by _____   _____
     POLITICAL SUBDIVISION   AUTHORIZED OFFICIAL   DATE

_____   by _____   _____
     POLITICAL SUBDIVISION   AUTHORIZED OFFICIAL   DATE
_____   by _____   _____
     POLITICAL SUBDIVISION   AUTHORIZED OFFICIAL   DATE

ACKNOWLEDGED AND AGREED BY THE DIVISION OF EMERGENCY MANAGEMENT

By: _____
            DIRECTOR

MUTUAL AID AGREEMENT
FOR EMERGENCY RESPONSE/RECOVERY
APPENDIX A

Date: _____

Name of Government: _____

Mailing Address: _____

City, State, Zip: _____

Authorized Representatives to Contact for Emergency Assistance:

Primary Representative

Name: _____

Title: _____

Address: _____

Day Phone: _____ Night Phone: _____

FAX No.: _____

1st Alternate Representative

Name: _____

Title: _____

Address: _____

Day Phone: _____ Night Phone: _____

2nd Alternate Representative

Name: _____

Title: _____

Address: _____

Day Phone: _____ Night Phone: _____

REQUIRED INFORMATION

Each request for assistance shall be accompanied by the following information, to the extent known:

1. General description of the damage sustained:

2. Identification of the emergency service function for which assistance is needed (e.g., fire, law enforcement, emergency medical, transportation, communications, public works and engineering, building, inspection, planning and information assistance, mass care, resource support, health and other medical services, search and rescue, etc.) and the particular type of assistance needed:

3. Identification of the public infrastructure system for which assistance is needed (e.g., sanitary sewer, potable water, streets, or storm water systems) and the type of work assistance needed:

4. The amount and type of personnel, equipment, materials, and supplies needed and a reasonable estimate of the length of time they will be needed:

REQUIRED INFORMATION (continued)

5. The need for sites, structures or buildings outside the Requesting Party's political subdivision to serve as relief centers or staging areas for incoming emergency goods and services:

6.   A specific time and place for a representative of the Requesting Party to meet the personnel and equipment of any Assisting Party.

ACKNOWLEDGMENT

To be completed by each Assisting Party.

NAME OF ASSISTING PARTY: _____

AUTHORIZED REPRESENTATIVE: _____

CONTACT NUMBER/PROCEDURES: _____

1. Assistance To Be Provided:

Resource Type	Amount	Assignment	Est. Time Arrival

2. Availability of Additional Resources:

3. Time Limitations, if any:

APPENDIX C

Diagram of Drill Scene

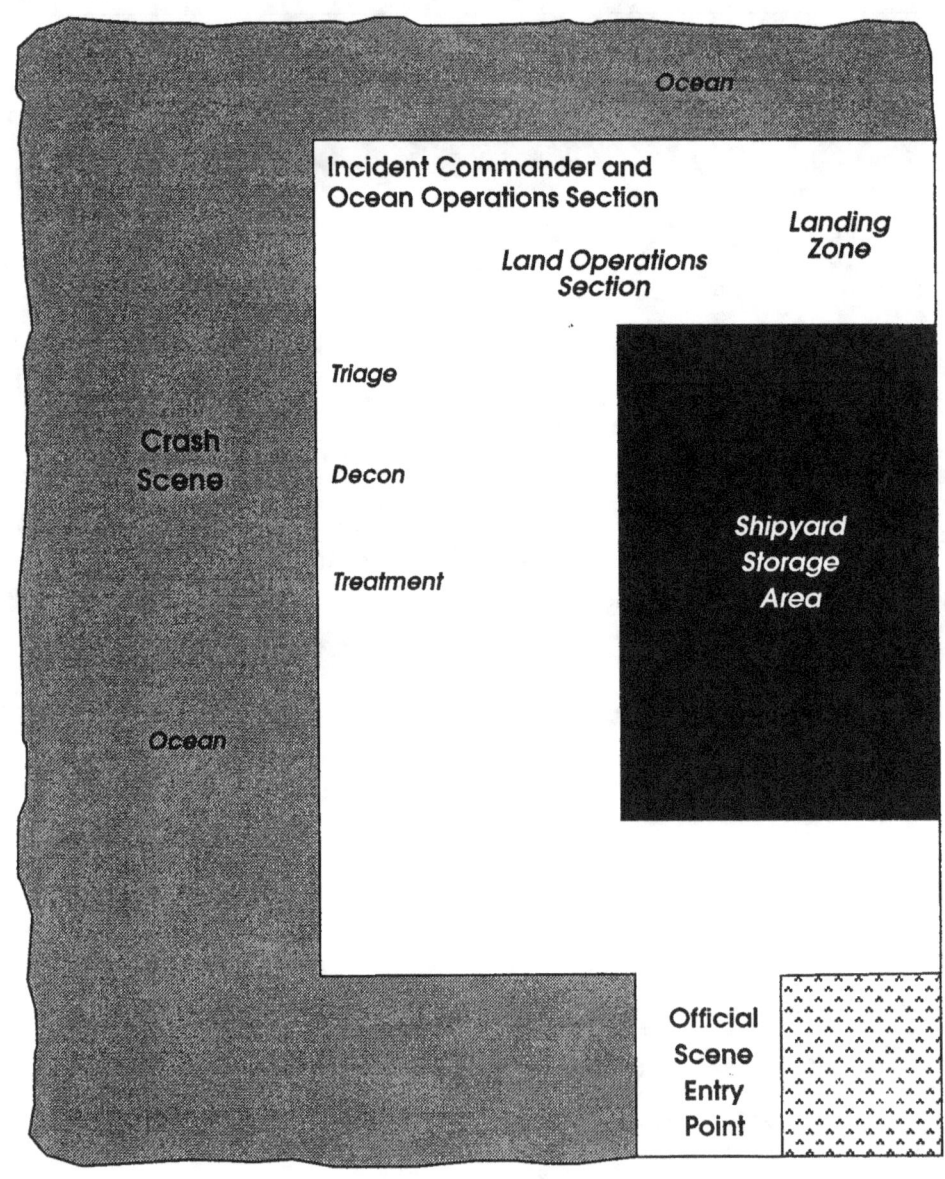

Drill Layout

3068-6-19-95-2

28

APPENDIX D

Photographs of the Drill

1. Each patient was given a tarp to lie on for drill purposes. In reality, multiple salvage tarps may be necessary instead.

Appendix D continued

2. The media provided excellent coverage of the drill but in some cases got in the way of operations, as pictured here.

Appendix D continued

3. Several mass casualty trailers with supplemental equipment were brought to the scene.

Appendix D continued

4. Signs were used to designate operations sections. The Land Operations Section comprised the Land Operations Section Chief as well as the Police Branch Officer and the Medical Branch Officer.

Appendix D continued

5. Victims were marshaled through the Triage, Decon, and Treatment areas here before moving on to the Transportation area. The signs helped maintain order and marked off each area in a manner that was easily visible to victims.

Appendix D continued

6. The Ocean Operations section was located on land but apart from the Land Operations sections. In reality, the Ocean Operations section may be established off-shore. At this drill, the Coast Guard Incident commander operated from the same location as the Ocean Operations section. Liaisons from each agency worked with the Incident Commander to facilitate operations and to provide inter-agency communications.

Appendix D continued

7. A mock floating fuselage made the drill scene realistic looking.

Appendix D continued

8. After the drill was complete, helicopters practiced removing victims from the water using Bill Pugh nets.

Appendix D continued

9. The local hazardous materials team established a decontamination area between the triage and treatment areas to simulate the removal of jet fuel and other contaminants.

Appendix D continued

10. Fire and rescue personnel from all agencies carried an ample supply of triage tags to the scene.

Appendix D continued

11. Helicopters from various agencies including the Coast Guard, Broward County Sheriff's office, and the Air Force were used to med-evac victims.